ISO Lesson (

I0069013

Pocket Guide to ISO 9001:2015

Fourth Edition

ISO Lesson Guide 2015

Pocket Guide to ISO 9001:2015

Fourth Edition

J. P. Russell

ASQ Quality Press
Milwaukee, WI

American Society for Quality, Quality Press, Milwaukee 53203
© 2016 by ASQ
All rights reserved. Published 2008
Printed in the United States of America
26 25 24 23 22   LS   8 7 6 5 4

Library of Congress Cataloging-in-Publication Data

Names: Russell, J. P. (James P.), 1945- author. | Arter, Dennis R., 1947-
    ISO lesson guide

Title: ISO lesson guide 2015: pocket guide to ISO 9001:2015 / J.P. Russell.

Description: Fourth edition. | Milwaukee, Wisconsin: ASQ Quality Press,
    2016. | Revised edition of: ISO lesson guide 2000: pocket guide to
    Q9001-2000 / Dennis Arter and J.P. Russell. 2001.

Identifiers: LCCN 2016023123 | ISBN 9780873899031 (alk. paper)

ISBN 9781636941318 (paperback)

Subjects: LCSH: Quality control—Standards—Handbooks, manuals, etc. |
    ISO 9001 Standard—Handbooks, manuals, etc.

Classification: LCC TS156 .A755 2016 | DDC 658.5/620218—dc23

LC record available at https://lccn.loc.gov/2016023123

ASQ advances individual, organizational, and community excellence
worldwide through learning, quality improvement, and knowledge
exchange.

Bookstores, wholesalers, schools, libraries, and organizations: Quality Press
books are available at quantity discounts with bulk purchases for business,
trade, or educational uses. For more information, please contact Quality
Press at 800-248-1946 or books@asq.org.

To place orders or to browse the full selection of Quality Press titles, visit our
website at: http://www.asq.org/quality-press.

Q
ASQ
Quality Press
600 N. Plankinton Ave.
Milwaukee, WI 53203-2914
Email: books@asq.org
Excellence Through Quality™

Table of Contents

Acknowledgement

Denis Arter is no longer an active contributor to this work but should be acknowledged for his past contributions. Denis Arter created an online paper about 9001 concepts in plain English in the late 1990s. I saw it and told Denis we could create a book with the material. He reluctantly agreed. After collaboration over the next several months we self-published the work as a little pocket guide. It was titled *ISO Lesson Guide: Pocket Guide to Q9001*, ©1998. Later in 2000, this same *ISO Lesson Guide* became ASQ Quality Press' first pocket guide.

Introduction

The *ISO Lesson Guide* translates ISO 9001 into easy-to-understand words. This pocket edition was designed as a quick reference for anyone to carry around conveniently.

Each element containing requirements is discussed and key concepts are highlighted at the beginning of each section.

Here is a rundown of the *ISO Lesson Guide* features:

- Quality is defined
- The ISO process approach is explained
- Key concepts are accompanied by an illustration
- Risk-based thinking is introduced
- Concepts are described in easy-to-understand words
- A brief conspectus summarizes ISO 9001 requirements
- Quality management principles are described in easy-to-understand words
- An entertaining fable explains the difference between ISO 9001 and ISO 9004

Ideal for handing out to existing and new employees, this pocket guide can also be used as supplemental study material for ISO 9001 training courses. Web-based training using the concepts in this pocket guide is available at www.QualityWBT.com.

Defining Quality

This pocket guide is a series of discussions on quality and the common methods used to achieve quality using the ISO 9001 standard. The first step is to define quality. That is not easy.

A popular book at the time I was going to college was *Zen and the Art of Motorcycle Maintenance*, by Robert Pirsig. Those of you who got past the first two chapters will recall that the hero was pursuing the definition of quality. It caused him to become mentally unbalanced. I hope that doesn't happen to you!

Let's take the simple approach in our definition of quality. Let's use the approach of ISO 9001, which is an extremely popular worldwide standard: Quality is what the customer says it is. Quality has another characteristic: It can be physically measured. If all the measurements add up to what the customer defined or expected, then what you just provided is quality.

Philip B. Crosby stated this many years ago in his book *Quality Is Free*. He defined quality as "conformance to requirements." Simple. The ISO 9001 standard builds on this concept of requirements for product, process, or system. Requirements either come directly from the customer or they come from you and are agreed to by the customer.

The second component of quality has to do with measurable characteristics. If something cannot be measured, it cannot possess quality. These measurements are physical and testable. This goes back to the military concept of form, fit, and function. Later on you will see just how important this concept of measurement is. If you can measure it, you can maintain it and improve it.

This approach implies that you could manufacture hat boxes that didn't hold hats (designer hat boxes) or video cameras without imaging devices (fake surveillance cameras). If you took steps to make sure that the physical characteristics of the said box or camera were defined and actually achieved, then you would have indeed made something of quality. If a hotel guest wanted to sleep on the floor and asked to have the bed removed, you would still be providing a quality service even though you were charging for a room with no bed. Everything must be seen from the eyes of the customer. Some products/services are fads. Remember Beanie Babies, Hula Hoops, Crocs, flare jeans, and Pet Rocks? Some products may be impractical or serve no practical function, but they were exactly what the customer wanted. This leads to the first rule of quality: Quality is defined by the customer.

Your customers know what they want, but you must also comply with the law of the land. Governments issue laws, ordinances, and regulations for the well-being, safety, and health of the public and to protect our environment.

Customer focus and improvement are critical when discussing quality. You have to provide quality or customers will take their business elsewhere. Sure, achievement of customer satisfaction is necessary, but to realize this higher level of performance, you must first have a command of the concepts. That's what this guide is about.

What Is a Process?

The quality management system standard takes the process approach. This simply means that the standard promotes organizing your work in the order in which you would normally do something. First you would plan it. Then you would do it. Then you would check and analyze what you did. Finally, you would improve on any weaknesses. The clauses of ISO 9001 can likewise be organized from start to finish, instead of as a random list of tasks. In our discussions you will see the letters PDCA. The letters stand for Plan-Do-Check-Act. The PDCA cycle is an old idea first published in the 1930s, yet we are still learning how powerful it can be.

A process is a series of steps or actions that do something. The something could be sawing, stamping, massaging, mixing, registering, or painting. A process represents action and usually ends with *-ing*. Henry Ford is well known for having lined up several processes to form an assembly line. Putting the processes in sequential order allowed Ford to provide affordable and reliable automobiles. When all the processes and assembly lines are put together, you have a system or organization that must be managed. For the best quality results, the system should be managed according to certain quality management principles (see page 92).

A system is a group of processes working together to achieve a common objective. The processes must be managed to achieve the organization's goals, such as meeting a budget or realizing a profit.

If you operate a farm, you may have cows that provide milk, chickens that provide eggs, equipment to plow the fields, a house to lodge workers, and a kitchen to provide meals. When you bring all this together you have a farm. The farm is the system supported by individual processes.

The following diagram shows the management system and processes that are part of this quality management standard (note the ISO 9001 clause numbers in the diagram).

Model for ISO 9001 quality management system (QMS)

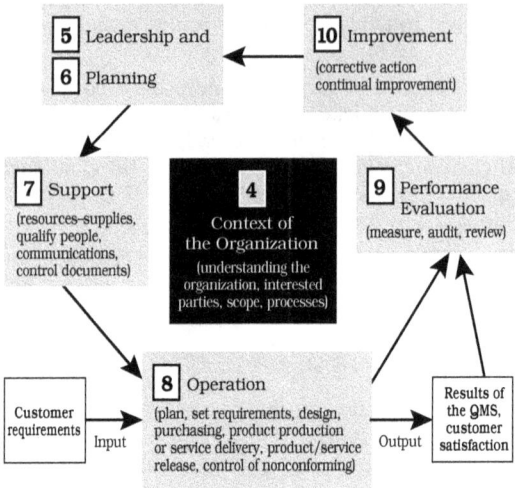

© 2015 QualityWBT Center for Education, LLC
Permission granted to reprint for non-commercial purposes

The important stuff really starts with clause 4 of the standard and continues through clause 10. Clauses 0 through 3 are introductory and administrative clauses that contain no requirements. You can see how the all-important customer fits into the system. The customer starts and ends the process. Your ultimate goal is to end up with a *satisfied* customer.

4 Context of the organization

4.1 Understanding the organization and its context

- *Describe what your organization does.*
- *List the important elements or factors for doing what you do.*

First, realize that *context* is a fancy word for description or profile. This is about describing or profiling what your organization does. It is similar to describing someone's personality. Are they independent or dependent on others? Flexible or inflexible? Private or sociable? Highly skilled or a generalist? And so on. Do you provide a product or perform a service? What good is it? What are key factors that make things work? (Supplier of specialty materials, easy credit, certified skills, and market linkages, regulatory instate, technology know-how, etc.) Next, what is your future? Is it the same old thing or do you have an idea or vision of something new? Sometimes it is hard to describe what you do and to think about your own future, but that is what ISO 9001 is asking organizations to do.

4.2 Understanding the needs and expectation of interested parties

- *List the persons or organizations interested in what you do.*
- *List why they are interested.*

Interested parties are people or organizations interested in what you do. Personally, this could include family members, your loan officer, the Internal Revenue Service, charitable organizations you donate to, and organizations you belong to. Some parties are not directly involved in your life but have an interest in it. In-laws might be a good example.

For your organization, interested parties are those that are not directly part of the customer-business relationship for the product or service you provide. Interested parties could include the local fire department, banks that provide financial resources, business partners, regulators, trade groups, unions, local governments, suppliers, and so on. Their needs or expectations could include: knowledge of hazards, safety issues, deposits, reports, information meetings, changes, and so on.

4.3 Determining the scope of the quality management system

- *Establish the quality management system boundaries.*
- *Write down what is covered by the quality management system.*

ISO 9001 is a set of rules. Does everyone need to follow the rules? Write down who has to follow the ISO 9001 rules and tell them. Do the ISO 9001 rules apply to all your organization's products and services? If you also produce a waste product or a byproduct as part of your manufacturing or service delivery, are they included? You may provide a primary service, but what about all the other services or information that you provide?

Are there ISO 9001 rules that do not apply or make sense for your organization? If so, note them and skip them.

4.4 Quality management system and its processes

- *Say what you do.*
- *Do what you say.*

Have you ever gone through a great deal of effort to figure out how to do a particular task? Have you ever had a key employee leave the company? Have you ever had goods returned because the tooling setups weren't quite right? These are all good reasons to develop a formal and structured approach to your operations.

You need to identify the important tasks and establish the sequence in which they should be performed. The *who, what, when, where* and *why* questions apply. Important tasks farmed out to others (outsourcing) must be included, too. When you are able to perform the processes correctly the first time around, you have an effective operation. You will need to identify measurements you will be taking and records you will retain as documented information so that you know how you are doing from time to time. This is the big-picture view of the quality management system (see the quality management system diagram on page 19).

Model for ISO 9001 quality management system (QMS)

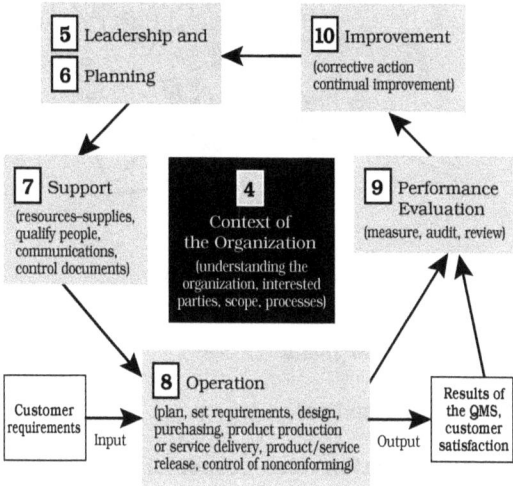

| 5 Leadership and |
| 6 Planning |

| 10 Improvement |
| (corrective action continual improvement) |

| 7 Support |
| (resources–supplies, qualify people, communications, control documents) |

| 4 Context of the Organization (understanding the organization, interested parties, scope, processes) |

| 9 Performance Evaluation |
| (measure, audit, review) |

| Customer requirements |
| Input |

| 8 Operation |
| (plan, set requirements, design, purchasing, product production or service delivery, product/service release, control of nonconforming) |
| Output |

| Results of the QMS, customer satisfaction |

5.1 Leadership and commitment

5.1.1 General

5.1.2 Customer focus

5.2 Policy (Quality)

5.3 Organization roles, responsibilities, and authorities

- *Provide the vision and show commitment.*
- *Define your policy on quality.*
- *Make assignments for providing quality products and services.*
- *Keep everyone informed.*

The ISO 9001 standard is called "Quality management systems—Requirements." It specifies important things you should do to assure your customer that you are providing what you promised.

Quality improvement is best when it starts from the top instead of from the lower levels within the organization. We want to do what the boss wants. Top management determines the vision, communicates the vision, provides the resources, and checks on the vision from time to time. For a small business, top management may be one person. For large organizations, top management may be the board of directors or the corporate officers. For most of us, top management is the boss, the person who has the last word on decisions at work.

You should start by defining quality as it applies to your business. This will lead to a policy on quality. While the policy itself is of little value, the act of defining it is of great value. Discussions among the members of your organization will lead to a clearer understanding of your business and your customers. You should keep your quality policy simple. A famous example out of California states, "We will sell no wine before its time."

After you have defined a policy on quality, you will need to organize the business to achieve that quality. The policy will lead to objectives, and the objectives will lead to a plan to achieve the objectives. Remember, planning is the first step in the PDCA cycle. Often planning is underrated, yet it may have the greatest value-added effect on the organization. When you go on a trip, you don't just get in the car or show up at the airport. You normally plan ahead. The next time you find yourself saying "That ran smoothly" or "That was easy," remember that someone or some group did a good job planning and developing the activity or process.

Of course, success factors other than customer satisfaction—such as safety, environmental, health, and financial controls—are also important. These may have their own plans or be integrated into an overall organization plan.

The next step is to define the duties and responsibilities of your organization. If you have more than one person attempting to do a job, then the roles and interaction of the various team members also must be defined.

Have you ever experienced the agony of not knowing what was expected of you in your job? How could you possibly do right by your employer when you didn't know what was expected of you? Deep down inside each of us is a basic desire to do the job well. If only we knew what that job was!

Even in a team environment, where there is much cross-participation, we still need to know the responsibilities of the various parties. While the phrase "Just Do It" may sell shoes, it cannot produce a quality product or service. Once all the team members have their assignments, keep them informed on how well they are doing.

Though not required, it may be best if one person is assigned overall responsibility and authority for quality. This person may be called the management representative, quality manager, director of quality, VP of risk, or some other title. You should know who this person is in your organization.

6 Planning

6.1 Actions to address risks and opportunities

- *If it might not work, what are you going to do about it?*

Now the standard asks that you consider the potential problems, issues, and concerns regarding how your organization operates (4.1) and what interested parties need (4.2). Technically, risk is the effect of uncertainty. Non-technically, risk could mean the potential problems or issues that will hinder or prevent you from doing what you want to do. On a personal level, what if you are going on vacation and the flight is cancelled? What if the bank does not get the payment on time? What are the consequences and what can you do about it?

For your organization, the risks are bigger. Perhaps you purchase an important material from a single source. What if they go out of business or there is a fire or hurricane? What will you do? In general, the idea is to have a plan for when things don't go the way you want them to. The same is true for interested party needs (4.2). What if the report is late or end users are not informed of safety issues? You can identify and assess risks through simple methods such as brainstorming, or use more formal methods if necessary.

You should analyze the importance of those possible problems to determine if the risk is acceptable. If the risk is unacceptable, you need to change how things are done. You can avoid the risk (stop doing what you're doing) or you can ease or mitigate the risk to lower the effect or possibility of occurrence. Perhaps the most common application of risk-based thinking is predictive maintenance of equipment. It can be as simple as thinking you need to use a timer to avoid burning food on the stove. You see, risk-based thinking is all about thinking.

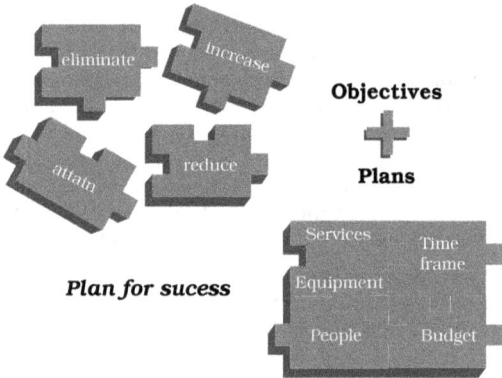

eliminate

increase

attain

reduce

Objectives

+

Plans

Plan for sucess

Services

Time frame

Equipment

People

Budget

6.2 Quality objectives and planning to achieve them

- *What are your goals and how do you plan on achieving them?*

You may have personal goals or objectives. These could include resolutions to get out of debt, ensure your kids graduate from high school, be to work on time, or take a vacation this year.

Your company/organization has objectives, too. They help move your organization in an intended direction. Sometimes they seem like the latest fad and go away in six months. Other times they make sense and become permanent. Objectives are important and it's important to write them down.

Setting goals is wonderful, but goals don't just happen. You need a plan to achieve your objectives. This requires some self-discipline.

6.3 Planning of changes

- *Know the reason for the change, address potential problems with the change, and don't mess up what was working.*

7 Support

Model for ISO 9001 quality management system (QMS)

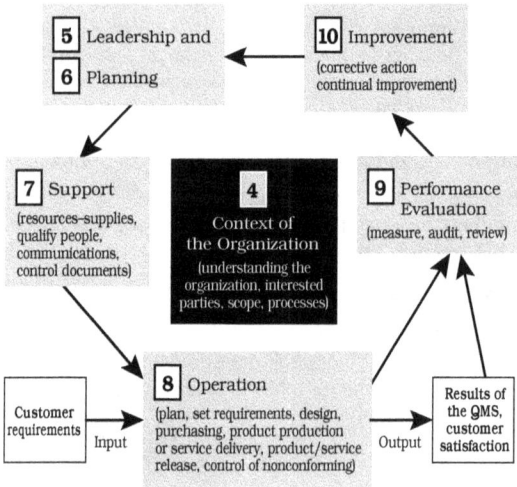

```
┌─────────────────────┐              ┌──────────────────────┐
│ [5] Leadership and  │      ←       │ [10] Improvement     │
│ [6] Planning        │              │ (corrective action   │
│                     │              │ continual improvement)│
└─────────────────────┘              └──────────────────────┘
        ↓                                       ↑
┌──────────────────┐  ┌─────────────────┐  ┌──────────────────┐
│ [7] Support      │  │      [4]        │  │ [9] Performance  │
│ (resources–      │  │  Context of     │  │ Evaluation       │
│ supplies,        │  │ the Organization│  │ (measure, audit, │
│ qualify people,  │  │ (understanding  │  │ review)          │
│ communications,  │  │ the organization│  │                  │
│ control documents)│  │ interested     │  │                  │
│                  │  │ parties, scope, │  │                  │
│                  │  │ processes)      │  │                  │
└──────────────────┘  └─────────────────┘  └──────────────────┘
        ↓                                       ↑
┌─────────────┐   ┌──────────────────────────┐   ┌─────────────┐
│ Customer    │ → │ [8] Operation            │ → │ Results of  │
│ requirements│   │ (plan, set requirements, │   │ the QMS,    │
│        Input│   │ design, purchasing,      │   │ customer    │
│             │   │ product production or    │Output satisfaction│
│             │   │ service delivery,        │   │             │
│             │   │ product/service release, │   │             │
│             │   │ control of nonconforming)│   │             │
└─────────────┘   └──────────────────────────┘   └─────────────┘
```

7.1 Resources

7.1.2 People

7.1.3 Infrastructure

7.1.4 Environment for the operation of processes

- *Provide resources for the management system (people–infrastructure–work place–equipment).*

Before you manufacture a product or provide a service, you first need some resources to accomplish the tasks. Resources include people, equipment, supplies, and facilities. For example: Before you start your baseball entertainment business, you need a field to play the game, a road to the field, bleachers for the spectators, a dugout or bench, a backstop, lights for night games, and perhaps a public address system. However, you have a limited amount of money to spend. Part of the fun of management is to find that balance between what you want and what you can afford and still satisfy the customer. Just like when purchasing a new car, you prefer the Lexus coup instead of the Ford subcompact, but you only need the car to drive to work and back. This is really part of the planning stage in the PDCA cycle: ensuring that you have the means to provide a product or service. You need to ensure there is a proper infrastructure before you begin production.

Resources include people to operate and control the processes. How many people do you need? Where will they come from? People need to be competent (see section 7.2) to operate the processes.

Equipment and facility needs can be internal or external to the organization. External resources may be needed due to specialized equipment requirements. Perhaps during initial phases of a project, purchasing resources outside the organization may not be justified.

People resources can also be internal or external. External resources may be used because of the need for specialized knowledge or skills. In some cases temp employees are used until full-time employees can be economically justified.

Once you place people in the facilities, you have a work environment. People work better and stay healthy when there are good working conditions relative to the task. If the task is grinding metal or shoveling coal into a furnace, you will never be able to keep the work area spotless. However, there should be clean break areas, masks to filter air the worker breathes, free fluids to combat dehydration, showers, ear protection, and so on. For the baseball field you may need restrooms or water fountains. For a large office complex you may need exercise facilities or ATMs. Look at everything you need to provide a proper work environment for the people performing the processes. Everything may include: a clean and safe environment; proper noise and vibration control; adequate lighting, temperature, and humidity control; and employee accommodations such as exercise rooms, stop-smoking programs, access to GED classes, and crisis counseling.

7.1.5 *Monitoring and measuring resources*

- *Identify information needed for go/no-go decisions.*
- *Provide equipment capable of providing that information.*

- *Maintain the equipment and use in the proper environment.*
- *Periodically check the equipment to ensure it is still okay to use.*

Control of resources (equipment) for monitoring and measuring is an important part of a quality management system. It allows you to make the go/no-go decisions on conformance to requirements. It's reasonable to expect that you need good data upon which to base these quality decisions. The only way you can get good data is through the use of good equipment. Understand this. You need good data. This clause is about *data* and how you can get the data needed for decisions.

Once you determine the measurement needs for incoming, in-process, and final product, install equipment that will give you those data. In addition to dials and gauges and spectrophotometers and tachometers, this also means the software running the devices. Make sure the equipment is set up correctly and calibrated. A common term for this is *grooming*. Make sure the equipment can do the job for you. Many service organizations need to verify the quality of equipment used to perform the service.

Once you have your equipment in good shape, you need to operate it in the environment for which it was designed. If the equipment is sensitive to temperature and humidity, don't use it outside in the rain!

Keep the equipment in good shape through maintenance and periodic checking. Often these devices will drift over time, so that the data obtained are no longer within your accuracy and precision bounds. Sometimes this can happen in a year, and sometimes it can happen in a couple of hours. Consult the instructions provided by the manufacturer of the equipment.

When it is important to be able to trace back a measurement to a device and time, some type of calibration program should be put in place. For example if a car air bag fails, you want to be able to trace back to the manufacturer, to who approved it, and to test results. The calibration status (such as next calibration check due date) needs to be known to the user. When you get an oil change, you get a sticker indicating mileage and when the next oil change is due. It is easy to check to see if you are overdue getting an oil change.

Sometimes you may find that your equipment is no longer reading correctly. This throws into question the quality of everything it measured since it was last declared to be good. While there is no requirement to always keep track of everything measured by each device, you do need to assess the damage or potential for damage. Remember, these resources were used to determine if you met the requirements for the job. That assurance, back to the customer, is important.

Some of you might be thinking, "Well, if this measurement stuff is so great, why not apply these concepts to all the instruments I use in my plant or facility? Why not apply calibration to in-plant controlling devices, as well as the go/no-go instruments?" While the thought is good, sometimes you can have too much control. The effort is just not worth the cost. Do you really want to trace the calibration of your steam pipe thermometers all the way back to the prime standard in Washington, DC? Is it of value to check the calibration of the drain flow meters? Here's where you need to make value judgments.

In summary, you need data to make the go/no-go delivery decisions called for in inspection planning. Good instruments will give you those needed data.

7.1.6 Organizational knowledge

- *Save job know-how.*
- *Gather knowledge that will be needed in the future.*

Sometimes the knowledge and skills acquired over years of practice is an organization's most valuable asset. You might call it know-how. Organizations may try to protect this know-how by classifying it as proprietary information. It is not exactly top secret, but it needs to be safeguarded. Examples of proprietary information include the formula for Coke or Colonel Sanders KFC recipe. Many organizations harbor special knowledge and skills that they do not want known, especially by their competition.

This clause is about properly documenting or recording process knowledge and skills. The organization needs to be able to pass on knowledge and skills to future operators of the process.

Some employees have acquired job knowledge and techniques that make things work like they are supposed to. In many cases, this has not been documented. Their contribution should be recognized and they should be asked explain what they do or document it. Some folks can show you what they do but cannot easily explain it.

The other aspect of organization knowledge is acquiring knowledge you need for the future. When the new machine arrives you will need the knowledge to operate it efficiently and safely. Launching a new service or product requires new knowledge to ensure objectives are met.

7.2 Competence

- *Prepare people so they can do the job.*
- *Ensure training and other actions were effective.*

You need to ensure you have a sufficient number of good people and that they know how to do the jobs assigned to them. However, no one is born with the knowledge to set up a DVD recorder, install an app, or use a cordless drill. Knowledge comes from the schools we attend and the books we read. Skills come from experience.

To know education needs and what kind of training to provide, you should perform a job needs assessment. Often when a person is first hired or during a performance review, the supervisor does a needs assessment. The supervisor (1) looks at the abilities of the employee, (2) reviews the projected resource needs of the group, and (3) meets with the employee to discuss the various options. A training and development plan can then be tailored so that the employee receives the training and experience needed to properly perform the job and qualify for future advancement. Sometimes the boss may decide that no additional training is needed for that individual. That's fine. The important thing is that competency needs are assessed. Often, the results of needs assessments are recorded in employee personnel files.

Once the training and development needs are determined, the next step is to fill those needs. Certain core classes may be desirable, such that all employees have an opportunity to participate. Sometimes training can be provided by internal employees. Sometimes it is cost-effective to look outside the company for training resources. These outside methods may include instructor-led courses, computer-based tutorials, or online (web-based) training programs. In all cases, you need to determine the effectiveness of the training provided.

Some tasks require special skills on the part of the employee. Certainly, auditors need to be certified. Inspectors also need to be qualified. Many process operators, such as welders and cleanroom assemblers, need to demonstrate specific skills. Process owners (audit boss, inspection supervisor, cell leader, field service manager, and so forth) should be the ones who determine which processes need people with special qualifications.

Of course, once you qualify an individual, you should record that qualification status in some sort of certificate. These can be general forms in the personnel records, certificates on the wall, or pocket cards. Have the process owners sign those certificates. Most folks are quite proud of their certification in a specific skill. If that pride can be built upon, everyone wins.

From time to time, check out the effectiveness of the training and other actions. Other actions may be mentoring, on-the-job skills training, exchange programs, testing, and so on.

Your goal is to have a quality management system operated by competent people. Competent people are those who know their job and know what to do when things are not as expected. Putting the right people in the right job is a good thing.

7.3 Awareness

- *Make sure people know what is expected of them.*

You might think *awareness* is an odd term for a quality standard versus *required job skills* or *training*. There are certain things you should know beyond operating the process(es).

You need to know your organization's policy on quality. I suggest you not try to memorize it, but instead be able to phrase the idea of it in your own terms to let people know you understand it. *We will always be prepared for every job. We only ship in-specification product... Conform to requirements... Make what the customer wants and get better,* etc.

The other things you need to be aware of are objectives, targets, or goals for your job relative to quality. If you get it right, how do you contribute to quality of the product or service? How do you and your organization benefit?

You also need to know what happens if you do not comply with the quality management system requirements. Could the job be shut down or the order be cancelled?

7.4 Communication

- *Plan what you're going to say.*

When the organization communicates, it must follow some basic rules related to what, when, to whom, how, and who communicates.

Communication is like the elephant in the room. It is often overlooked but can be the cause of many problems. Today we can communicate many ways, including cell phones, text messages, email messages, VoIP, intranet, video, etc. Emailing is great for one-way communication but not so good when you need to negotiate an issue.

What is perhaps the number one reason for not doing the job right? Yes, that's right: "No one told me" or "I did not know."

7.5 Documented information

7.5.1 General

7.5.2 Creating and updating (i.e., documented information)

- *Write down the important things.*
- *Get organized to achieve quality.*

Once you have the general direction laid out, it is time to plan for the important tasks that need to be done. Written plans and procedures are job performance aids. They are written for a trained individual, not a newcomer. Two to six pages are normally all you need. A procedure is used to define the major steps to be accomplished. It can be captured in any way that is convenient to the user: paper, computer file, table, or picture on the wall.

Often, detailed job specifications are needed to implement a general procedure. For example, general instructions are used to define the operation of a printing press, while the job spec sheet is used to state the paper type, ink color, and quantity to be run for a specific customer. Likewise, welders use a blueprint drawing to make a particular joint, and teachers use a lesson plan. These are all job-specific instructions for a general task.

You should create documents where you need them, to ensure effective control of the operation. The standard does not require any specific written procedures, so it is up to you to decide what documentation is needed.

A common foundation for many quality management system includes a manual. The manual, whether on the intranet, cloud, or three-ring binder, should capture the important tasks so that they are done correctly the first time around. Doing it right the first time is your quality contribution for your organization.

Start by having a team (six or fewer people representing interested groups) examine the very highest organizational document you have and then work down. The highest-level document might be a corporate policy or a master franchise agreement or a federal rule on processing juice. List the ten to twenty processes that are really important. Now write one descriptive paragraph on how you do each of those important processes in your organization. Keep the paragraphs short and to the point. Next, keeping in mind the PDCA cycle, draw a picture to show how the processes connect. When you bring all this together, for all the processes, you have a manual. The collection of documented information may be called an operation manual, service manual, business manual, quality manual, or other name you chose.

As you can see, the concept of having a quality management system is good. It defines your activities, assures quality products and services, and helps you in achieving your goals of customer satisfaction and never-ending improvement. Here, then, is the second rule for quality: Define the tasks and write them down.

7.5.3 Control of documented information

- *Make directions available to the users.*
- *Keep them up to date as long as you need them.*
- *Identify needed records and maintain them.*

Documented information control deals with the first item in the PDCA cycle: planning. But first, it is necessary to understand the difference between a document and a record. A document tells you what to do. A record tells you what you did. A document comes before the job, whereas a record comes after.

The quality management system is composed of generic concepts supported by local descriptions, which are supported by job procedures, which are supported by task details. All of these items are considered to be documents, in that they define the requirements for a task. These defining documents for the job need to be correct and current. So, it is necessary to have some method of approval for correctness before the documents are actually used. They should be approved to say, "All the desired information is included and we've decided that this is the approach we want to take." It is common to keep a record of this approval process, either on the document itself (such as a signature) or by some other record stored in a safe place.

After the document has been approved, it needs to get to where it is needed, such as to the users. What good is a grocery list when it's up on the refrigerator and you're standing in the dairy aisle? Often these books, manuals, and procedures are in the work area. In some companies, they are placed online electronically, where they can be reviewed when needed.

As in most human endeavors, we often don't get it exactly right the first time. Change is necessary to correct mistakes or incorporate new information.

Since you went to a lot of trouble to generate these documents and get them distributed, you certainly don't want to mess them up through the change process. So, it is necessary to have approvers (or someone with their knowledge) make the changes or check the changes before they are issued. It is also necessary to make sure that the version being used is current. This is called configuration control.

Sometimes you may have documents from outside the organization such as equipment operation manual or customer specifications or approved industry methods. These documents must be controlled too. The level of control is up to you.

Records are an important part of any quality program. They allow you to review prior activities for present application. They also provide assurance to others that people are doing what they promised.

Records are one of the most believable forms of communication.

Records need to be controlled similar to documents. Before you start collecting records, you must define the records to be kept. Rather than attempting to generate a whole list of records to cover every potential operation of your organization, you should identify the records in the individual operating (process) procedures. For each procedure, state three things: what records are to be generated by this operation, how long those records are to be kept, and where they will be kept. Your records are now much more manageable. Once you have defined the records to be kept, keep them in good shape.

Records can be useful tools for managing quality. Like hot sauces, though, they need to be used in moderation. The key to good records management is to truly understand why a particular piece of information is needed and how best to capture that information. Records are very powerful and give you a greater understanding of processes. They help managers make decisions. They also help others understand your operations. Make the best use of this powerful tool.

8 Operation

Model for ISO 9001 quality management system (QMS)

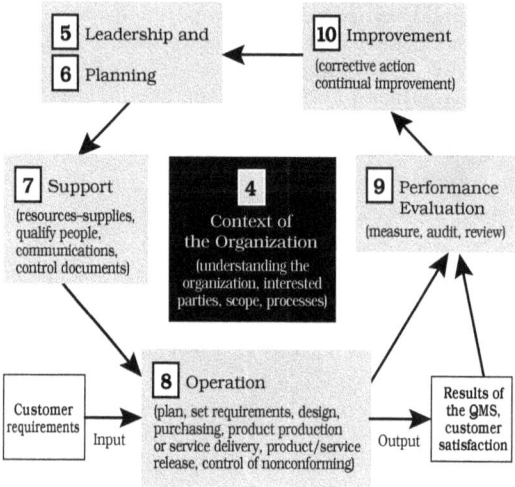

| 5 | Leadership and |
| 6 | Planning |

10 Improvement
(corrective action
continual improvement)

7 Support
(resources–supplies,
qualify people,
communications,
control documents)

4
Context of
the Organization
(understanding the
organization, interested
parties, scope, processes)

9 Performance
Evaluation
(measure, audit, review)

Customer
requirements

Input

8 Operation
(plan, set requirements, design,
purchasing, product production
or service delivery, product/service
release, control of nonconforming)

Output

Results of
the QMS,
customer
satisfaction

8.1 Operational planning and control

- *Determine the process steps ahead of time.*
- *Decide on the documents and records you will need to confirm process outputs are as planned.*

The word *operational* (ready to operate) in the title means to manufacture a product or provide a service. Earlier you defined a plan of the overall system; now you need to plan the processes, from receipt of the order to final delivery of the product or service. Determine the important process steps, checks, records, and goals. This may be a flowchart or process description. Most people like pictures because they are easy to understand. Pictures enable people to better understand the process linkages and identify potential risks or opportunities.

8.2 Requirements for products and services

8.2.1 Customer communication

- *Clearly understand the customer requirements for the product or service.*
- *Make sure you can do it.*
- *Keep the customers informed and listen to them.*

Remember the definition of quality? Conformance to requirements. This element of the ISO 9001 standard is extremely important to your goal of producing a quality product or service and achieving customer satisfaction. In fact, your quality program depends on your ability to define those quality characteristics.

The pioneers of the quality movement defined these characteristics as form, fit, and function. In other words, these quality characteristics consist of physical things that can be measured and tested. Size, color, texture, chemistry, and weight are all examples of quality characteristics. Services also have measurable characteristics such as on-time delivery, clean bed linens, and correct information such as telephone numbers.

The concepts in clause 8.2 need to be applied to whatever method you have for determining your customer requirements. Sometimes it can be as simple as choosing Number 12 from the restaurant menu. Perhaps you have a catalog listing all the products you provide. All the customer has to do is state the catalog item number and the quantity desired. In this case, you have specified the physical characteristics and the customer has agreed to those parameters. It may also be prudent to agree to contingency plans if there could be delays or requirements can't be met.

Sometimes it is not so easy. A landscaping contractor has general parameters that can be described in a brochure, but the specifics must be discussed and agreed to with the customer. A fruit packing firm must determine not only the type of fruit, but also the size and color ranges acceptable to the purchaser. The packing company may also be requested to supply only organically grown fruit, individually wrapped in tissue. A tire company may not only provide tires meeting specifications but also be responsible for the disposal of worn or defective tires. A misunderstanding here could result in millions of dollars in losses by the time the shipment arrives at its destination.

An extension of this process of defining customer requirements is to make sure that those requirements and subsequent changes actually get to your employees. The sales crew can have a wonderful program for defining the order in great detail, but if the packing clerk doesn't get the information, the customer is stuck with something unusable. Another example might be the automotive service shop. The counter representative may have actually seen the location of the oil drip under your car, but if that information doesn't get to the mechanic, the problem may not get resolved.

Figure out the best way to communicate with the customer—from the order to the delivery process— by fax, e-mail, internet, or telephone, for example. While keeping customers up to date, listen for changes in their expectations.

There are times a customer may provide material to be used to produce a product or part. Or they may provide special packaging. In either case, the organization must agree to how the customer-owned property is handled and controlled. Perhaps you experienced a contractor using and damaging one of your tools to perform a job at your home. Did they replace it with a new one?

8.3 Design and development of products and services

- *Create a design plan.*
- *Know what you are designing.*
- *Identify measures for success.*
- *Review the work as it progresses.*
- *Verify that you did what you promised.*
- *Make sure it actually works.*
- *Scrutinize changes.*

This is classic design stuff. It was originally written to cover the situation where the customer has asked you to invent something new. If your customers knew what they wanted, they wouldn't be asking you for design help. Additionally, the current standard requires design and development controls anytime it is necessary to provide product that meets customer and regulatory requirements. Simply, if your organization is responsible for design, there should be design controls. When you provide a service, such as conducting an advertising campaign for a new line of cookies or designing training programs, you will certainly need to apply these design principles.

First, you need to assemble and equip a design team and develop a plan. Then, you get the client or sponsor to agree to this plan. After all, it is their money you are spending! The plan is not a static document. It will need revision as the project progresses.

After the plan is developed (and approved), you now need to understand the design requirements. How many football players must the locker room hold at peak capacity? How fast do you want to water the lawn? Will the brochure be two or twenty pages? Many studies have shown that the main cause of project failure is an inadequate understanding of the design input requirements going into the project.

Once you have defined the functional requirements of the product, you need to define the product or service itself, including those things that can be measured to show success. Now the actual creative process can take place. Do not let the design proceed too long without some intermediate checks. Typically, design reviews are held early on (at perhaps 20 percent completion), midway through (around 50 percent), and when the project is nearly finished (about 90 percent). These reviews are intended to critically examine the progress against all requirements.

Before handing the finished design over to the client, make sure you did what you promised in the design plan. (Remember that?) Common practice is to hold a final design review and publish a report listing those promised items checked, along with the results. It is sort of like a final inspection of products leaving the factory. Additionally, designs may be verified by independent checks, alternate calculation, prototype testing, and other means. The verification step lowers risk and builds team confidence in the design.

Before you can call it quits, you need to do one more important thing: Make sure the whole thing works. The client has hired you because of your talent. Sure, the thing could be exactly what the client specified, but if it doesn't work, then it is not a quality job. We call this final task "validation." In many applications (software, for example) there is still a lot of magic mixed in with experience.

You can see that these concepts of design have direct application when you are providing a creative service. You don't need to be an architect to understand their importance. You just have to modify the classic phrase "The customer is always right" to "I have to help the customer achieve what he or she expects."

8.4 Control of externally provided processes, products, and services (i.e., purchasing)

- *Know what you want from outside providers/organizations.*
- *Okay them and check them out from time to time.*

Most of the time it is necessary to buy material or services from others to operate your business or make your product or deliver your service. Perhaps you don't have the skills, knowledge, or equipment to do a portion of the contracted job, so you decide to subcontract that portion to someone else. It's still your neck on the line, however. If the lamp switches in your manufactured product wear out after a month, does the customer complain to you or to your switch supplier? You, of course. You can explain to the customer that it was the switch manufacturer's fault, but the customer will still be dissatisfied.

Just as clause 8.2 of ISO 9001 stresses the importance of understanding your customers' requirements (remember measurable characteristics), it is equally important for you to define your requirements to your supplier or other external providers of products or services. There are actually three items that need to be defined here: the technical requirements for the job or product, the criteria you will use to accept the work, and the quality system you expect to be followed when doing the work. Normally these items are spelled out in a purchase order or contract.

Once you have defined your requirements, you then need to find and select a supplier. Shop around. Your selection will depend on a number of parameters such as price, delivery date, reputation in the marketplace, and the supplier's ability to meet your defined requirements. Once you've selected your suppliers, you mustn't forget them. If the contract is a lengthy one (typically more than a year), you will need to formally evaluate the supplier's performance. This can be done by reviewing its delivery history, talking to the users of its stuff, or visiting its plant or facilitates. This periodic review is becoming especially important as organizations move toward partnerships and away from strictly awarding business to the low bidder.

It is not uncommon to read about a business that has gone under because it couldn't get good parts and components. Usually the problems lie with both the buyer and the seller. If you take the time to clearly state your requirements and then see if the other folks can truly meet those requirements, you may succeed where others have failed.

8.5 Production and service provision

8.5.1 Control of production and service provision

- *Control your processes to ensure good outputs.*
- *When you cannot check the product, check the process.*

This is the most important section in the entire ISO 9001 quality management system standard! The application of process controls will determine your success or failure as an organization/business. Sure, you can deliver goods that have been thoroughly inspected and tested. You can exchange those items that manage to get through with defects. You can repeat the service or offer discounts. But you won't stay competitive very long when you have high internal costs and dissatisfied customers. The exception is when there is very little competition and that leaves end users little choice. For example, the lowest American Customer Satisfaction Index (April 2016) ratings go to subscription television service, internet service providers, the post office, government services, and health insurance.

To start, you must first identify your product and service processes. A great way to do this is by flowcharting. List the ten or fewer steps it takes to accomplish your business. Keep each statement on this list short; for example, "Do something." Now draw a box around each statement and arrange the steps in the order accomplished. Finally, connect the boxes with lines going in and out. The entire flowchart should take up no more than a page.

Now you need to carry out those processes under controlled conditions. For each of the processes, you should (1) provide acceptance criteria for a job well done, (2) have work instructions, pictures, or notes on how to perform that process or task, (3) make sure the process and equipment are capable of doing what you want, (4) monitor and adjust critical parameters of the process, and (5) implement follow-through processes for release, delivery, technical service, and after-sales service. While these concepts of process control seem to represent good common sense, they are often lacking. Some processes are performed by robots where work instructions have been converted to software programs.

Please understand that you can get by with less than desirable process control and still keep the customer happy. For wrong services, you may respond with an aggressive refund, discount, or upgrade program. For wrong product, a strong inspection program will identify errors and mistakes before the customer even sees the final goods. But in either case, you will have high internal costs and employee frustration. You also won't be competitive with businesses that learned how to do it right the first time.

An important aspect of controlling a process is when outputs cannot be verified by subsequent tests for conformance. If you cannot check something before the customer receives it, you must validate the process that produced it. An example would be cooking soup inside a sealed can. You don't know if the cooking process was correct unless you open the can. Now the soup is no longer sealed and cannot be sold. Other examples of special processes include welding, heat treating, soldering, and crimping. Virtually all services are received by the customer as they are performed. The quality of a service cannot be checked before it is performed, therefore service process training should be validated and verified it is effective.

For these special processes, you may need to qualify the process, the equipment, and the operator. In many cases specified procedures requiring certain records must be used. Revalidation may be required periodically or when the process is changed.

8.5.2 *Identification and traceability*

- *Match the specs to the job.*
- *Show whether items are acceptable or not.*
- *Keep track of what you provide.*

As previously mentioned, the ISO 9001 quality management system standard defines quality as conformance to requirements. As items progress through the various steps of processing, you need to know if they are ready for the next operation. This makes sense. It also keeps you from performing an action that might prevent later analysis of the item's quality. Historically, a common method used to accomplish this control is to have a sheet of paper accompany the item as it travels through the operation. That's where we got the word "traveler" for the sheet of paper. The inspector marks the traveler sheet to show that an item has passed or failed inspection. When the traveler card has all the inspection results listed (as satisfactory, of course!), the product is ready to ship. Today, an organization may identify a product with a bar code, scanners keep track of its progress, and status is updated via software.

Another way to know the status of a product is to brand the item or assign a lot number. When we buy underwear, we must peel off the Inspector 15 sticker before wearing it. An electronic component may have quite a few inspection stickers inside the cabinet to indicate satisfactory inspection. Sometimes the inspector is issued a stamp (also called a "bug") to show that the piece, pipe, or component has passed. The results of software testing are often recorded directly into the code so that other programmers can see the needed information. Now, most of the steps can be done electronically via software.

This is one of the earliest forms of quality control. It allows your fellow workers to know the quality of something before they work on it. More important, it allows customers to know the quality before they buy it. This is an easy one to implement.

Another concept contained in this section of the standard is traceability. This could mean one of several things, depending on your customer (or regulatory) requirements. One form of traceability is to be able to match the parts and materials to the finished item. This is very useful for failure analysis. It is called backward traceability.

Another form of traceability calls for the identification of the finished product by stamp or serial number. If you have ever activated a cell phone, you will need the serial number. Those product registration cards you fill out when you purchase a new microwave ask for a serial number. This allows the manufacturer to trace the sale back to the factory and date of production. An even tighter form of traceability is practiced in the food processing and pharmaceutical industries. Here, the manufacturer needs to be able to trace the distribution of the product in case of recall. This is called forward traceability.

The ISO 9001 quality management system standard does not require any form of traceability. Those requirements come from your customer or your regulator. For many businesses, traceability is not necessary.

Product identification, status, and traceability are concepts that have stood the test of time. They are nothing special. They are good business practice.

8.5.3 *Property belonging to customers or external providers*

- *Don't break other people's stuff.*

The first question you must ask is, "Do I have any customer property or stuff that belongs to someone else?" You may be using equipment that belongs to someone else, an external provider. A customer may give you stuff to make their product or provide the service, and the customer may want it back—in good shape, thank you very much. Customer property can include intellectual property such as methods, formulas, or drawings. Also, it can include personal data that must be safeguarded such as credit card information. The key here is ownership. If the customer retains ownership while the property is in your possession, you should take a closer look. A good example would be dresses left at a dry cleaning establishment. The customer wants them back. They are temporarily in your possession. Another example would be empty containers supplied by the customer and filled by you. Finally, how about those broken or virus-infected computers that you have been given to repair? Or the car that has been left at your repair shop? These are also customer-supplied items. You certainly do not want to return them in worse shape than you received them.

Upon deciding that you do, indeed, have customer or other property not belonging to you, you first want to inspect it as it comes in. Note the condition and try to get the customer or provider to acknowledge that condition. Now, put the stuff in an appropriate place. (Don't leave it out in the backyard in the rain.) If it needs periodic cleaning, greasing, or wiping, then you are obligated to perform this maintenance while the item is in your possession.

What if you break it? While you certainly don't want that to happen, you must own up to it if it does. The ISO 9001 quality standard requires you to record the damage and notify the customer or other external provider. It doesn't tell you how to resolve the problem you now have. That's between you and your customer or other external provider.

The concept of protecting customer property includes intellectual property that may be confidential or proprietary information contained in documents. It includes private information stored and recorded on a computer. Private personal information includes financial transactions and medical history.

This portion of the ISO 9001 quality management standard is good business. It is also good for society. It's the right and proper thing to do.

8.5.4 Preservation

- *Keep good stuff good.*

You have gone through a lot of trouble to get good material. You have controlled your processes. You have inspected items to make sure they conform. And now you throw all of it into the warehouse and forget about it? Items need to stay in a state of goodness as they sit around waiting to be used or shipped. This may include preserving packaging, labels, or product security.

If you produce golf balls, they probably do not need much attention for several months. After all, they are designed to perform well outside, even after getting whacked by a big stick! On the other hand, microchips need care and attention before they are stored in their protective (and grounded) boxes. Apples must go into cold storage before they turn brown. So, examine your inspection operations. See if the items passing inspection need attention prior to their next inspection.

Handling and storage must be appropriate for the item being held. Here's where the concept of shelf life appears. Some items will deteriorate with age. Reagents in a laboratory need to be used within a specified period of time. Food will spoil if it is left out on the counter. Even if they have no specified shelf life, many items need special care while they are in storage. Winter coats may become moth-infested over the summer. Beans will sprout if they get wet. Steel tanks will rust if left outside in the rain.

Some items need attention while they wait in storage. Motors are a good example. If the assembly sits in one position over a lengthy period, it actually begins to bow. Bearings start to corrode. The solution is to rotate that machine every so often. Electronic equipment is adversely affected by moisture, so desiccant crystals are placed in the shipping box.

Some companies apply special controls to the shipping of their products to the customer site. While this has always been a consideration in the selection of shippers, it has not traditionally received a great deal of attention. We are beginning to realize that customers don't care who is at fault when they receive damaged goods. They just get mad! That unhappiness will often cause them to select another supplier for their next purchase. It is economical for you and your shipper to examine the shipping process.

For some services it may be important to preserve certain aspects or elements of a service so it can be performed. The service may require certain equipment or access to the internet, cabling, or security software that needs to be preserved.

8.5.5 Post-delivery activities

- *Don't bail on the customer.*

In many cases, when you ship the product or provide the service, it ends your responsibilities, but other times not. The product you provide may need servicing, have disposal requirements, need to be upgraded, and so on. If you have post-delivery responsibilities, do it with a smile.

8.5.6 Control of changes

- *Be careful when making changes.*

For improvement, there must be change, but not all change results in improvement. Encourage change but be cautious. Thoroughly discuss changes and let everyone know about it. As with the corrective action process, test your changes before they go system wide.

8.6 Release of products and services

- *Check the product against requirements.*
- *Ship only what was ordered.*
- *Write down if it was good and who said so.*

The third rule of quality is "Do it right the first time." The item or task must meet requirements. It is far too expensive to rework, scrap, or sort product after the fact. Quality for the supplier is doing it right the first time. The processes for providing a product or service must be measured against some acceptance criteria. Organizations should know when processes are operating as intended. For example: Is the process the right temperature, pressure, level, amperage, rpm, rate, and sequence? If the production and delivery processes are performing properly, they are likely to result in an acceptable product and customer satisfaction.

Product quality is defined as conformance to requirements. We have learned that these are measurable and testable features, specified by the customer (or stated by you and accepted by the customer). Quality for the customer is getting what they were promised or expected.

We now need to discuss the ways in which you can assure your customer that you are shipping quality items and providing quality service. You know that the raw materials and parts going into your product are important. You should verify that they meet your standards before you use them. Just as a good cook uses only the freshest ingredients, you need to check the quality of what you are using. Tests and inspections are sometimes appropriate. Supplier auditing can also be useful. Paperwork provided with the material is often helpful. The point is that you know what you are using.

Inspection and testing during production or delivery may also be useful. Go back to the flowchart of your operations that you prepared for process control (clause 8.1 of the ISO 9001 standard). Identify those points where the cost and risk of nonconformance are greater than the cost and risk of inspection. Of course, work should not proceed past these points until everything is found to be acceptable.

A long-accepted method of assuring the quality of your work is to inspect it before you ship it. This is referred to as final inspection (or test). All previous inspections and tests should be completed (and accepted) before the product is shipped. Paperwork must also be in order. In case of subsequent failure or dispute, these inspection records can be quite valuable. Services cannot be inspected before they are performed, but products used to provide a service can be inspected prior to use.

This is classic go or no-go decision. In the past, operators could not inspect their own work. They had to get QC to sign off on the job. This approach, although still acceptable, is not required. Many firms are now placing the responsibility for inspection directly in production.

Inspection and testing will give you the data to show yourself and the customer that you are producing to requirements. It is noble and will always be necessary. However, the coverage and intensity of inspection will decrease as control of processes increases.

8.7 Control of nonconforming outputs

- *Keep bad stuff away from good stuff.*
- *Figure out what to do with the bad stuff.*

First of all, let's define what we mean by non-conforming. Again, we need to go back to the definition of quality: conformance to requirements. So, how do you know if something conforms to requirements? You test it! Either it passes or it fails. There is no middle ground here. So, the definition of nonconforming product is something that flunked a test. From earlier discussion, we know that something can be tested as incoming raw material, in-process components, or finished goods.

This concept of pass or fail is an important one for understanding the control of nonconformances. When you draw a sample of spaghetti sauce from the pan and taste it, you are analyzing the process (of cooking). The data (from your taste) will tell you if the process needs adjusting (more garlic!). This is not a nonconforming situation. You, the operator, will adjust the process. It's no big deal.

However, if the spaghetti sauce flunks final inspection at supper, it is a big deal. Your family (the customers) will tell you what to do with it. You are not allowed to make that decision. You have five basic options: rework, repair, regrade, reject, or release. These are called remedial actions.

Rework means to take it back to the stove and get it into a state of conformance. In the process industries, there is a variation of this called recycle (blend or dilute). If you squirt the bad stuff into the process stream slowly, it will mix sufficiently to form an acceptable product. (Sometimes there is strength in numbers!) Repair means to make it acceptable, but not as originally specified. Perhaps you can pick out the burnt ground beef and substitute clams instead. It's edible, but it's not exactly as originally specified. To regrade means to offer it as it turned out, but at a lower grade (and price). You seldom break even here. To reject means to dump the whole pot of sauce down the garbage disposal. Even the dog wouldn't eat that stuff! The last possible disposition is to tell everyone you messed up and release it: "Why, sure, it tastes funny, but we all love you anyway."

There are also interim actions when you are not sure what to do with bad stuff. They call this containing, segregating, or holding. Eventually the "on-hold" stuff must be dealt with.

Services can be nonconforming, too. Materials or equipment to be used in a service could be found to be nonconforming prior to the service being conducted. Perhaps the scissors were dull or the wheel balancer was defective. There can be after-the-fact nonconformities when the customer or end user tells you about it, such as the spaghetti sauce. Perhaps you cut the person's hair too short. What next? If customers experience nonconformities, they may not complain but find another supplier.

Control of nonconforming product and service is a big deal. Rarely should the operator be allowed to determine the disposition. The hassle factor is quite high. This is as it should be. You want these failures to hurt! Only then will they receive the attention necessary to prevent repetition. Workarounds will only cause your company to be less efficient and thus less profitable. It reminds us of the saying "no pain, no gain."

9 Performance evaluation

Model for ISO 9001 quality management system (QMS)

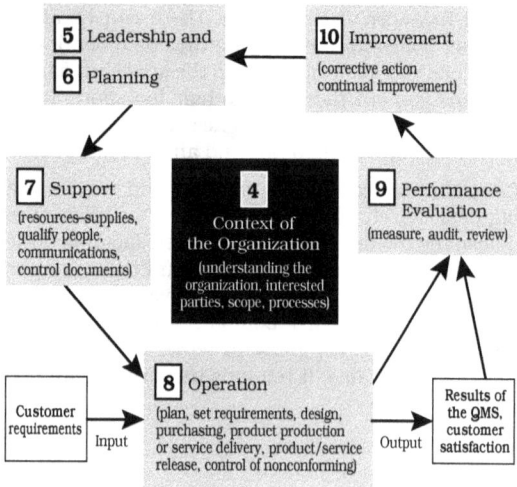

© 2015 QualityWBT Center for Education, LLC
Permission granted to reprint for non-commercial purposes

9.1 Monitoring, measurement, analysis, and evaluation

9.1.1 General

9.1.2 Customer satisfaction

9.1.3 *Analysis and evaluation*

- *Identify important process measurements.*
- *Collect data and decide when it needs to be evaluated.*
- *Keep track of customer satisfaction.*
- *Figure out what the information means.*
- *Use data to make sure things are right and when to make things better.*

Think about the information you need to collect and how you will use it. Ask yourself if you are collecting the right data and if it is useful. Many organizations keep data that have no value, while others fail to analyze data that could provide important insights for improvement. Get a picture of your information plan by identifying the sources of information, how information is collected, and where information goes. Information may come from measuring product characteristics, process performance, audit results, complaint reports, surveys, and so on. Some organizations find it useful to flowchart the collection, use, analysis, and evaluation of information.

You need to decide on the *what, how,* and *when.* For example, you may decide that you need to measure your weight because your objective is to keep your weight in a certain range. To answer how, you decide to use a digital scale and record the results. The when requires you to weigh each morning after breakfast. Next, you decide to analyze the data once a month to determine if your weight is trending up or down.

Recall rule #1: The customer defines quality. Regardless of the tight tolerances or speed of delivery, if customers are not happy, they may seek out other suppliers. You may link customer satisfaction to conformance to customer requirements, customer responses such as complaints, and customer perception of your product and organization. On a company's balance sheet is a section called assets. There is a space for both tangible and intangible assets. Tangible assets are physical things like a building, an automobile, and the stuff stored in the back room. Intangible assets are things like brand-name recognition, market position, customer goodwill, and loyalty. Customer loyalty is related to customer satisfaction and has value. If you wanted to purchase a baseball team, the tangible assets would only add up to a few bats, baseballs, and mitts. The value of the baseball team would be in the intangible asset column. You need to gather information about tangibles as well as the intangibles to provide a measure of customer satisfaction.

Customer satisfaction is about providing customers with what they asked for at a fair market price, while treating them with respect. It is about customer perception of quality received, price for the value, and how they were treated. Customer satisfaction may also be market and individual-dependent. It could depend on an individual's past experiences. Some customers have higher expectations than others. The level of customer service your competitors provide may also affect satisfaction.

Studies show it is more economical to spend money to keep current customers happy than to constantly seek new customers through sales and marketing efforts.

In clause 9.1 you created a map (flowchart) to identify sources of information, how it is collected, and where it goes. Here you need to determine what data you are going to analyze and evaluate to effectively measure performance and identify areas that need to be improved. This is the brain of your quality management system. The output of this clause will provide factual information that can be used to assess progress and make management decisions. Evidence-based decision making is one of the seven Quality Management Principles.

You should collect and analyze all the data that will allow you to determine system effectiveness and identify areas for improvement. The standard writers wanted you to be sure to provide information on the following areas: (1) customer satisfaction, (2) conformance to requirements, (3) performance, (4) results of actions to address risks and opportunities, and (5) need for improvement.

Some of the most commonly used tools for analysis and evaluation of data include the seven basic quality tools:

- Flowcharts
- Cause-and-effect diagrams
- Control charts
- Histograms
- Check sheets
- Pareto charts
- Scatter diagrams

You have gathered data. You have plotted those data on nice charts. Now you need to analyze and evaluate those charts. What are the data telling you? A common problem is to miss the trends because you see the same set of data every day.

An occasional review of those charts and graphs by outsiders and auditors will increase your chances of detecting subtle changes. This review also gives you added assurance that your processes are stable and that your measurements of product or service characteristics are valid.

Study the possibility of applying statistical techniques to your business. You may decide that these additional controls are nice, but you can't afford them. Fine. You are off the hook. You may decide that nothing in your operations would benefit from statistical techniques. That's fine, too. Or you may decide that the application of statistical techniques would benefit your operations and you want to use them. Great!

The study of which statistical techniques to use and where to use them is normally done by people knowledgeable in the use of statistics. Of course, you would want input from everyone involved in the process when performing this study. Some of you may already be using statistical process control, or SPC, in your operations. Do you know why? Where is that study?

Once your study is done, you need to start implementing these statistical techniques. Rather than write separate procedures, you are generally better off incorporating these ideas into your standard operating procedures.

9.2 Internal audit

- *Examine your internal operations against requirements.*
- *Report the results to those in charge.*
- *Ensure action is taken to fix problems.*

There are four very critical controls specified by the ISO 9001 standard for quality management systems: quality system (define your methods to achieve quality), process control (control your work operations), internal audits (examine your operations), and corrective action (eliminate the causes of problems). These four items also relate to the concepts of Plan-Do-Check-Act. As you can see, auditing is very important.

The first thing needed is an audit schedule. Normally covering a year's worth of audits, the schedule identifies what areas are to be examined and when.

The schedule should take into consideration the most important areas, recent changes, and prior audit results. The audit boss should publish the audit schedule. It should also reflect the desires of those senior managers in charge of the areas to be audited. Although there is no requirement to do so, most organizations plan their audits so that every part of the standard is examined within a period of one year.

Auditors may be drawn from the resources of the company. They should be able to conduct the audit in an impartial and objective manner. If they own the process, they cannot objectively analyze that operation. They need to be trained in quality auditing techniques. This may be accomplished through classes, reading, or observing others. Auditors also need a technical understanding of the processes being audited. (You cannot audit an operation unless you can first flowchart that operation.) As you can see, auditors are skilled individuals.

Audits always examine the operations against certain standards of performance. Performance standards include this ISO 9001 standard and the documents defined in clause 7.5. Generally these criteria consist of the standard operating procedures developed for the job. They can also include drawings and manufacturers' technical manuals. Often, internal audits will only judge conformance to these criteria. The more mature audit programs also judge the effectiveness of those criteria. As you might imagine, assessment of effectiveness is hard to do.

Audit results are published in the form of reports. They tell the managers how the operations are being performed. There is no requirement to limit the report to only bad things. Good implementation of controls should also be acknowledged. An audit report should contain four things: background on why the audit was conducted and what was examined, summary of the application of the controls examined, any specific good conclusions, and any specific bad conclusions.

Once the report is published, nonconformances need to be addressed and causes of significant problems need to be eliminated. The group audited needs to develop an action plan to fix the problem(s).

Many organizations have procedures and forms specifically for taking action to address audit results. The forms are commonly called corrective action requests (CAR) and they allow individual problems to be identified, analyzed, corrected, and tracked to completion. When this type of program is in place, many organizations automatically fill out a CAR form for each adverse audit finding. Then the audit may be closed and the problems will still receive the attention they need.

9.3 Management review

- *Measure your progress.*
- *Make improvements.*

You have defined your quality policy, planned for its implementation, identified responsibilities, and defined authority to achieve your goals. Now it is necessary to periodically measure your progress and take steps to continually improve. We call this management review. It can be done by your own people or by outsiders. The period of time between reviews varies from organization to organization. The review should be formal and needs to look at the bigger picture of your operations. "How are all the processes of our business working to achieve the goal of superior quality and customer satisfaction?" is a fair question. Typical items to be reviewed might be identification of internal problems, customer complaints, customer satisfaction survey results, employee turnover, new and repeat orders, and supplier performance. The outcome of your review should be a determination of the need to change the system. Just as human bodies need periodic medical checkups that may result in a change in a person's habits, businesses also need examinations to determine their health and their future direction.

10 Improvement

Model for ISO 9001 quality management system (QMS)

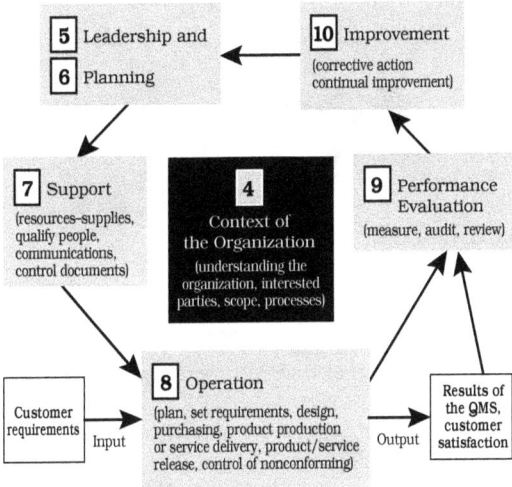

5 Leadership and
6 Planning

10 Improvement
(corrective action
continual improvement)

7 Support
(resources-supplies,
qualify people,
communications,
control documents)

4
Context of
the Organization
(understanding the
organization, interested
parties, scope, processes)

9 Performance
Evaluation
(measure, audit, review)

Customer
requirements
Input

8 Operation
(plan, set requirements, design,
purchasing, product production
or service delivery, product/service
release, control of nonconforming)
Output

Results of
the QMS,
customer
satisfaction

10.1 General

10.2 Nonconformity and corrective action

10.3 Continual improvement

- *Practice never-ending improvement.*
- *Identify problems.*
- *Determine why the problem occurred.*
- *Fix the cause of the problem.*
- *Verify that your changes worked.*
- *Be open to new ideas that would improve how things are done.*

Controls give us confidence that objectives will be achieved. But the world is not perfect, and controls sometimes don't work. Mistakes happen. Usually, you fix the mistake and life goes on. Sometimes, the mistake is (or could be) serious. People are getting hurt, people could lose their jobs. Internal costs are rising. Something needs to be changed. The whole system needs attention. That's why this concept of corrective action is so hard to implement.

Before you can change the system, you need to know which problems need your attention. Just as an athlete needs to conserve energy for the big event, you need to focus your attention on the big problems. The small items will be resolved during the normal course of doing business. The trick is to figure out the difference between big and small items. Although each organization is different, three things to consider might be cost, opportunity, and risk. If you don't fix the problem, what are your out-of-pocket costs, what efficiencies or improving capacity opportunities are being missed, and what customer loyalty or government intervention risks are you accepting? You need to determine the kick-in criteria that work for you. Too low a threshold will cause you to look upon everything as a crisis. Too high a threshold will allow serious problems to continue and eventually shut you down!

Corrective action is about eliminating the cause(s) of a problem so that it does not happen again. Corrective action should be reserved for heavy-duty, serious (known) problems that significantly affect cost, opportunity, or risk. Once you have declared a problem to be serious, you need to determine why it happened. This is where you apply all those wonderful problem-solving tools, such as flowcharting, brainstorming, Six Sigma, and so forth. Underlying cause analysis usually takes more than a day and should involve several minds.

After you have discovered the underlying cause of the problem, you need to fix it. We are not talking about the problem itself, but rather the cause. Very seldom is this an easy thing to do. (You have already solved the easy problems, after all!) You need help from all the organizational elements of the group here. The solution can seldom be accomplished within 30 days. Capital equipment may need to be purchased. New employees may be needed. An action plan is required, along with commitment dates and responsibilities.

OK. You have taken the action and changed the system. After a period of time (normally between six months and a year), you need to see if your great plans actually worked. Has the problem recurred? Has the chance of occurrence diminished? Are folks happier with the new controls? Revisiting the problem is an important but often overlooked part of corrective action.

Expect to hear a lot about corrective action. It is the part of your quality management system that makes things better for you and your organization. Corrective action and risk-based thinking are the fuel for practicing never-ending improvement.

> *Final thought: The quality management system controls are only as good as the integrity of the people responsible for maintaining them.*
>
> – JP Russell

ISO 9001 Conspectus

The ISO 9001:2015 standard for quality contains good business practices and is based on sound quality management principles. Here is a concise description (conspectus) of all the elements:

Model for ISO 9001 quality management system (QMS)

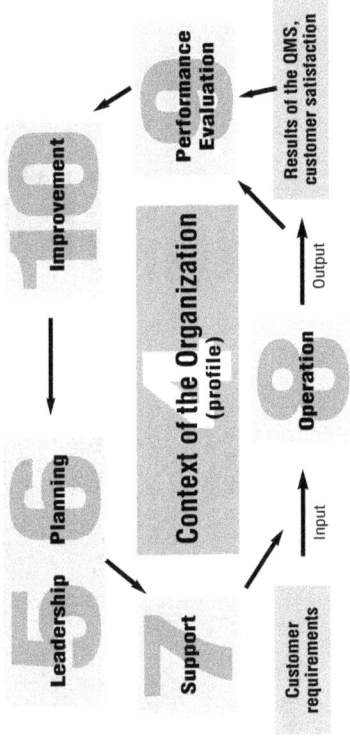

Leadership **5**

Planning **6**

Improvement **10**

Performance Evaluation **9**

Results of the QMS, customer satisfaction

Support **7**

Context of the Organization (profile) **4**

Operation **8**

Output

Input

Customer requirements

4 **Context of the organization**	4.1 Understanding the organization and its context	• Describe what your organization does. • List the important elements or factors for doing what you do.
	4.2 Understanding the needs and expectation of interested parties	• List the persons or organizations interested in what you do. • List why they are interested.
	4.3 Determining the scope of the quality management system	• Establish the quality management system boundaries. • Write down what is covered by the QMS.
	4.4 Quality management system and its processes	• Say what you do. • Do what you say.

5
Leadership
	5.1 Leadership and commitment
	  5.1.1 General
	  5.1.2 Customer focus
	5.2 Policy (Quality)
	5.3 Organization roles, responsibilities, and authorities
		• Provide the vision and show commitment.
		• Define your policy on quality.
		• Make assignments for providing quality products and services.
		• Keep everyone informed.

6 Planning	6.1 Actions to address risks and opportunities	• If it might not work, what are you going to do about it?
	6.2 Quality objectives and planning to achieve them	• What are your goals and how do you plan on achieving them?
	6.3 Planning of changes	• Know the reason for making changes, address potential problems with the changes, and don't mess up what is working.
7 Support	7.1 Resources	• Provide resources for the management system (people–infrastructure–work place–equipment).
	7.1.5 Monitoring and measuring resources	• Identify information needed for go/ no-go decisions. • Provide equipment capable of providing that information. • Maintain the equipment and use in the proper environment. • Periodically check the equipment to ensure it is still okay to use.

7 Support (continued)	7.1.6 Organizational knowledge	• Save job know-how. • Gather knowledge that will be needed in the future.
	7.2 Competence	• Prepare people so they can do the job. • Ensure training and other actions were effective.
	7.3 Awareness	• Make sure people know what is expected of them.
	7.4 Communication	• Plan what you're going to say.
	7.5 Documented information	• Write down the important things. • Get organized to achieve quality. • Make directions available to users. • Keep them up-to-date as long as you need them. • Identify needed records and maintain them.

8
Operation

8.1 Operational planning and control

- *Determine the process steps ahead of time.*
- *Decide on the documents and records you will need to confirm process outputs are as planned.*

8.2 Requirements for products and services
  8.2.1 Customer communications

- *Clearly understand the customer requirements for the product or service.*
- *Make sure you can do it.*
- *Keep the customers informed and listen to them.*

8.3 Design and development of products and services

- *Create a design plan.*
- *Know what you are designing.*
- *Identify measures for success.*
- *Review the work as it progresses.*
- *Verify that you did what you promised.*
- *Make sure it actually works.*
- *Scrutinize changes.*

8.4 Control of externally provided processes, products and services (i.e., purchasing)

- *Know what you want from outside providers/organizations.*
- *Okay them and check them out from time to time.*

8.5 Production and service provision	• Control your processes to ensure good outputs.
	• When you cannot check the product, check the process.
8.5.1 Control of production and service provision	
8.5.2 Identification and traceability	• Match the specs to the job.
	• Show whether items are acceptable or not.
	• Keep track of what you provide.
8.5.3 Property belonging to customers or external providers	• Don't break other people's stuff.
8.5.4 Preservation	• Keep good stuff good.
8.5.5 Post-delivery activities	• Don't bail on the customer.
8.5.6 Control of changes	• Be careful when making changes.
8.6 Release of products and services	• Check the product against requirements.
	• Ship only what was ordered.
	• Write down if it was good and who said so.
8.7 Control of nonconforming outputs	• Keep bad stuff away from good stuff.
	• Figure out what to do with the bad stuff.

9 Performance evaluation	9.1 Monitoring, measurement, analysis and evaluation	• Identify important process measurements. • Collect data and decide when it needs to be evaluated. • Keep track of customer satisfaction. • Figure out what the information means. • Use data to make sure things are right and when to make things better.
	9.2 Internal audit	• Examine your internal operations against requirements. • Report the results to those in charge. • Ensure action is taken to fix problems.
	9.3 Management review	• Measure your progress. • Make improvements.
10 Improvement	10.1 General 10.2 Nonconformity and corrective action 10.3 Continual improvement	• Practice never-ending improvement. • Identify problems. • Determine why the problem occurred. • Fix the cause of the problem. • Verify that your changes worked. • Be open to new ideas that would improve how things are done.

Quality Management Principles

1. **Customer focus:** Take great pride in meeting customer requirements; don't reward yourself for following procedures. You depend on customers, so you should understand their requirements and strive to exceed their expectations.

2. **Leadership:** Continual improvement and customer satisfaction are achievable when led by top management. Leaders provide direction and make sure everyone is on course.

3. **Engagement of people:** People, not machines, make quality a reality. When vision, objectives, and plans are shared, everyone will be working together to benefit the organization.

4. **Process approach:** Organize work the way it naturally flows. When activities are linked together, there is a structure for effectively managing and improving. Grouping processes creates a system. Top management organizes the processes into a system to meet the requirements of customers and interested parties.

5. **Improvement:** To achieve excellence, constantly look for ways to make things better. Make any changes in a controlled manner. Verify you got what you expected.

6. **Evidence-based decision making:** Make decisions based on the facts. Don't assume or be fooled by perceptions; analyze and evaluate information, then decide. There is a reason for everything.

7. **Relationship management:** Treat suppliers as business partners, not as servants. Your performance and reputation depend on suppliers and other interested parties, so work toward win-win outcomes.

A Fable

Here is a short story about the difference between ISO 9004 and ISO 9001.

Once upon a time, in a kingdom far away, things were not going well. Pestilence, plague, famine—these were everywhere. Those inhabitants who could were leaving for greener kingdoms. Those who couldn't leave were barely surviving, unable to deal with any of the more complicated issues in their lives. The king was starting to feel quite uncomfortable. This had never happened to him before.

This king was a good king. He took a lot of things for granted, though, and hadn't bothered to keep up with changing conditions.

As his father had done before him, the king asked the wizard what he should do to amend the situation. The wizard had recently read something from across the seas and it impressed him greatly. He said to the king, "We should use ISO 9004. It is a quality management system guideline standard for helping organizations be successful." Wow! What a powerful, yet simple, statement!

The king started to define the "interested parties." (This was a new term to him.) Once he had these defined, he asked what they wanted. Often the answer was simple, like "Water for the crops." Sometimes the wants of the interested parties were more difficult, and more than the king could provide.

Gradually, however, things got better! The subjects started to smile. The kingdom became well known. Other countries began to covet conditions in this kingdom. The king liked this.

Life in the kingdom continued to improve, but it reached a point where small changes no longer had large results. The knights studied the situation and concluded that most of the difficulties that remained did not come from within the kingdom. The problems came from the suppliers outside the kingdom. "We have our act together," reported the knights, "but those folks in the surrounding kingdoms just don't get it."

Once again, the king went to the wizard. The king asked, "What can we do to get our suppliers to give us what we need? This variation in grain moisture is causing us fits. The butter we get is often rancid. The timbers are too big, and they aren't straight, either."

The wizard went back to his books and studied. He came once more to the king and said, "ISO 9001. The name for this is 'Quality management systems—Requirements.' We know what we want. We now need to make sure that our suppliers give us what we specify."

A proclamation went out to all of the surrounding suppliers. If they wanted to do business with this king (and who didn't?), they had to agree to practice the new science of Quality management systems—Requirements.

Some suppliers mumbled, "We don't need no stinkin' QMS." They were quietly dropped from the list.

Others said, "This is really hard stuff, but we think it is worth it." Sure enough, their quality got better. The king was happy. He ordered more supplies from them.

The king trusted his wizard and he trusted his knights, but he still didn't trust his suppliers. Sure, some kept their promises, but others began to slide.

Once more, the king asked the wizard what he should do. The wizard thought and thought. How could this kingdom be sure that those suppliers were doing what they promised? Sure, the king could send his knights out to the far kingdoms to keep an eye on them. The knights, however, were worn out from their recent crusades. Besides, other kingdoms didn't particularly like being watched by these outsiders.

Then it came to him. Why not ask the knights from the kingdom of Certification Bodies, a legion well-respected far and wide, to check on the suppliers? They were known to be pure of heart and trustworthy. The wizard went to the king. He said, "We should have our suppliers checked out by the knights of Certification Bodies. This is called 'Quality Management System Certification.'" He went on to say, "As long as the suppliers remain in good standing with those knights, we should continue to give them our business."

The king agreed with his wizard and said, "Make it so."

The kingdom continued to prosper. Peasants came from across the seas. They settled in, learned the language of Juran and Deming, and raised families. The king was much exalted and the kingdom grew very wealthy. Everyone lived happily ever after.

NOTES

NOTES

NOTES

NOTES

www.ingramcontent.com/pod-product-compliance
Lightning Source LLC
Chambersburg PA
CBHW070356200326
41518CB00012B/2255